SURVEY OF HIGHER EDUCATION INFO TECHNOLOGY LEADERSHIP: USE OF ZOOM

AF875442

ISBN 978-1-57440-641-2 ©2020 Primary Research

Contents

THE QUESTIONNAIRE

1. Address

   A. Name
   B. College Name
   C. Work Title
   D. Country
   E. Email Address

2. Is your college public or private?

   A. Public
   B. Private

3. What is the full time equivalent enrollment of your college or university?

4. What is your institution's type or Carnegie class?

   A. Community College
   B. 4-Year College
   C. MA/PHD Granting
   D. Research University

5. What is the average full year tuition at your institution prior to discounts or student aid?

6. About how much of your work time over the past month would you say has been largely devoted to issues concerning the functionality of Zoom?

7. As a personal estimate about what percentage of your school's online classes utilize zoom?

8. How would you rate the general capabilities of your college faculty in using zoom?

   A. Overall most are highly proficient
   B. Most are generally proficient
   C. Most are proficient but many need help
   D. Most still need help though some are proficient
   E. Most need a lot of help and only few are proficient

9. How would you rate the general capabilities of your college students in using Zoom?

   A. Overall most are highly proficient
   B. Most are generally proficient
   C. Most are proficient but many need help
   D. Most still need help though some are proficient
   E. Most need a lot of help and only few are proficient

10. Which of the following features do you find faculty often need help within Zoom? (check all that apply)

    A. Setting up passcodes
    B. Using polls
    C. Recording and transcripts
    D. Managing student access and admission
    E. Screen sharing
    F. Chat functions
    G. Whiteboard
    H. Calendars
    I. Breakout rooms
    J. Security features
    K. Other (please specify)

11. Has your school developed videos or tutorials to help in learning Zoom?

    A. Yes
    B. No
    C. No but in the process of making them

12. Please write for your peers a short thirty second address on exactly what to avoid in Zoom and exactly what do to use it to its greatest potential

13. What has been your most effective tutorial or other tool to bring instructors and students up to speed in Zoom?

14. Which programs or applications do you think best complement Zoom for college classes?

15. Once the pandemic concludes and either through a vaccine, treatment, superb contact tracing or herd immunity, college life returns to something approaching normal, how do you think colleges will utilize Zoom?

16. Over the past two months how serious have the security problems been with your college's use of Zoom?

    A. Extremely concerning
    B. Very concerning
    C. Somewhat concerning
    D. Not so concerning
    E. Not at all concerning

17. What advice do you give faculty and students about protecting Zoom sessions from intrusion or hacking?

18. What are some of the very basic things that faculty and students tend not to know about Zoom?

PARTICIPANTS LIST

Alverno College
Arizona State University
Arkansas State University-Mid South
Central Oregon Community College
College of the North Atlantic Qatar
Eastern Washington University
EPFL
Foothill-De Anza Community College District
Grand Rapids Community College
Hagerstown Community College
Hobart and William Smith Colleges
Indiana University of Pennsylvania
Lansing Community College
Liberty University
Linn-Benton Community College
McGill University
Midlands Tech
Monroe Community College
Mt. San Antonio College
Oregon State University
Princeton University
Roosevelt University
Sacred Heart University
Saint Xavier University
South College
Southeast Mo State University
UNC Charlotte
United Lutheran Seminary
University of British Columbia
University of Hawaii, Manoa
University of Kentucky
University of South Dakota
University of Toronto
University of Wisconsin - Eau Claire
Wagner College
Walden University
Wilkes University

Characteristics of the Sample

Is your college public or private?

	#
Public	26
Private	12
Total	38

Enrollment

	#
Less than 4000	9
4000 - 9999	8
10000 - 20000	10
More than 20000	11
Total	38

Carnegie Class or type of college

	#
Community College	11
4-Year College	6
MA/PHD Granting	8
Research University	13
Total	38

Tuition, $

	#
Less than 5000	10
5000 - 10000	8
10001 - 20000	11
More than 20000	9
Total	38

SUMMARY OF MAIN FINDINGS

Nature of the Sample

We interviewed a sample of 38 information technology personnel drawn from centers of teaching and learning, information technology departments, instructional design support services and academic libraries. The most common titles of those interviewed were: CIO, instructional designer, IT Support, instructional technologist, and professor. Many were high level officials in information technology departments. The surveying was conducted in the last two weeks of May, 2020. The sample included 11 from community colleges, 13 from research universities, 6 from BA-level colleges and 8 from MA/Doctoral level universities.

Percentage of Work Time Devoted to Zoom Functionality

We asked: About how much of your work time over the past month would you say has been largely devoted to issues concerning the functionality of Zoom?

The average amount of work time spent on Zoom was 23.41% with a median of 10% and a range of 0 to 100%. Those working for 4-year colleges were by far spending the most time, a mean of 85% in the sample.

Percentage of Classes that Utilize Zoom

We asked: As a personal estimate about what percentage of your school's online classes utilize Zoom?

The mean was 61.53% with a median of 67.5% and a range of 5% to 100%. Use was dispersed relatively equally by size and type of institution though 4-year colleges were using Zoom the most – 81% of their classes utilized Zoom, while community colleges utilized it the least, 46% of their classes involved Zoom use.

Rating College Faculty Skills in Zoom

We asked: How would you rate the general capabilities of your college faculty in using Zoom?

We gave five possible answers to this multiple choice question which were: Overall most are highly proficient Most are generally proficient Most are proficient but many need help Most still need help though some are proficient Most need a lot of help and only few are proficient

A plurality of 42.11% answered "Most are generally proficient" while the second most common answer – given by 31.58% -- was "Most are proficient but many need help." Personnel in larger public colleges and in community colleges were more likely than others to feel that their constituency needed more help with Zoom. All of those feeling that their clientele need a lot of help and that only a few were proficient were working in community or 4-year colleges rather than in more research-oriented institutions.

Rating the Zoom Skills of Students

We asked: How would you rate the general capabilities of your college student in using Zoom. Once again, we gave five possible choices, the same as those for the prior question.

A plurality of 50% chose "Most are generally proficient" and the second most common answer given by 26.32% of those sampled was "Most are proficient but many need help."

Zoom Features that Faculty Need Help With

We asked: Which of the following features do you find faculty often need help with? The first we asked about was setting up passcodes. 18.42% though that this was something faculty often needed help with; all of those who thought so worked at public colleges and all were from community colleges or research universities.

Help with Use of Polls with Zoom

Next, we asked about use of polls, and 39.47% thought faculty often needed help with this. The larger the college the more likely it was for IT personnel to think that faculty needed help in this area. Fully 54.55% of those from colleges with more than 20,000 students felt that faculty needed help in this area.

Help with Use of Recording and Transcripts in Zoom

Next, we asked about recording and transcripts in Zoom and 39.47% of respondents felt that faculty needed help in this area.

Help with Managing Student Access and Admissions in Zoom

We asked about managing student access and admission in Zoom; 36.84% felt that faculty needed help in this area, including 42.31% of public college respondents. Nearly 64% of community college respondents felt that their faculty had issues in this area.

Help with Managing Screen Sharing in Zoom

We asked about screen sharing in Zoom. A full 50% of respondents said that faculty need help in this area; those from public colleges were almost twice as likely as those from private colleges to feel that faculty needed help with screen sharing in Zoom.

Help with Managing Chat Functions in Zoom

In contrast only 18.42% felt that faculty needed help with chat functions in Zoom.

Help with Managing Whiteboards in Zoom

The use of whiteboards in Zoom was an area that 55.26% of those surveyed felt that faculty needed help. Nearly 77% of respondents at research universities felt that faculty needed help in this area.

Help with Managing Calendars in Zoom

Only 10.53% felt that faculty needed help with use of Calendars in Zoom.

Help with Managing Breakout Rooms in Zoom

68.42% felt that faculty needed help with breakout rooms in Zoom, including nearly 82% of community college respondents.

Help with Managing Security Features of Zoom

44.74% felt that their faculty needed help with the security features of Zoom.

Development of Video Tutorials for Zoom

We asked: Has your school developed videos or tutorials to help in learning Zoom?

42.11% said yes; and equal number, no; 2.63% were in the process of making them. Half of all public colleges have made them; a quarter of private colleges have done the same as have 63.64% of community college respondents.

Advice on Using Zoom to its Greatest Potential

We asked: Please write for your peers a short thirty second address on exactly what to avoid in Zoom and exactly what do to use it to its greatest potential.

In general, respondents provided a very broad range of pointers and advice for their peers. Some basic advice includes enjoinders to use reminders before starting classes, encourage faculty to understand first what they want to do functionally before learning specific Zoom features, use waiting rooms and passcodes and much else.

Most Effective Zoom Tutorials

We asked: What has been your most effective tutorial or other tool to bring instructors and students up to speed in Zoom?

Some mentioned videos available from Zoom; others, to their own efforts, still others to third party sites sources such as: California ConferZoom.

Best Complementary Programs for Zoom

We asked: Which programs or applications do you think best complement Zoom for college classes?

For the most part the learning management system, though Canvas seemed preferred. Turning Point, Google Docs and Slack were also mentioned.

Likely Path of College Use of Zoom after the Pandemic

Once the pandemic concludes and either through a vaccine, treatment, superb contact tracing or herd immunity, college life returns to something approaching normal, how do you think colleges will utilize Zoom?

For the most part respondents felt that Zoom was here to stay, if not to the same degree. A general consensus might be that it will certainly continue to be used for meetings and online classes but that it would also encroach on more traditional classes, perhaps even encouraging some hybridization. Nonetheless, as might be expected, there was much disagreement on the likely future use of the technology. However, its impact is likely to be felt profoundly even as the virus (presumably) passes.

View of Security Problems with Zoom

We asked: Over the past two months how serious have the security problems been with your college's use of Zoom?

We gave five possible answers to this multiple choice question, which were: Extremely concerning Very concerning Somewhat concerning Not so concerning Not at all concerning

18.42% of respondents did not answer this question but a plurality of 36.84% chose "not so concerning" and 15.79% each chose "somewhat concerning" or "not at all concerning."

Private colleges were somewhat more likely than their public college counterparts to feel that security issues were "extremely" concerning.

Advice on Protecting Zoom Sessions from Hacking or other Intrusions

We asked: What advice do you give faculty and students about protecting Zoom sessions from intrusion or hacking?

In general, respondents felt that any security issues could be controlled with a little prudence and foresight.

We asked: What are some of the very basic things that faculty and students tend not to know about Zoom?

Respondents had a lot to say and mentioned many features of potential use, too numerous to list.

Table 1.1 About how much of your work time over the past month would you say has been largely devoted to issues concerning the functionality of Zoom?

	Mean	Median	Minimum	Maximum
Entire sample	23.41	10.00	0.00	100.00

Table 1.2 About how much of your work time over the past month would you say has been largely devoted to issues concerning the functionality of Zoom? Broken out by your college is public or private?

Is your college public or private?	Mean	Median	Minimum	Maximum
Public	23.60	10.00	0.00	100.00
Private	22.86	10.00	0.00	70.00

Table 1.3 About how much of your work time over the past month would you say has been largely devoted to issues concerning the functionality of Zoom? Broken out by enrollment

Enrollment	Mean	Median	Minimum	Maximum
Less than 4000	17.67	7.50	0.00	70.00
4000 - 9999	18.33	15.00	5.00	40.00
10000 - 20000	40.13	17.50	5.00	100.00
More than 20000	13.57	5.00	0.00	50.00

Table 1.4 About how much of your work time over the past month would you say has been largely devoted to issues concerning the functionality of Zoom? Broken out by Carnegie Class or type of college

Carnegie Class or type of college	Mean	Median	Minimum	Maximum
Community College	18.25	7.50	0.00	90.00
4-Year College	85.00	85.00	70.00	100.00
MA/PHD Granting	11.83	8.00	0.00	40.00
Research University	22.27	20.00	0.00	75.00

Table 1.5 About how much of your work time over the past month would you say has been largely devoted to issues concerning the functionality of Zoom? Broken out by tuition, $

Tuition, $	Mean	Median	Minimum	Maximum
Less than 5000	15.75	3.00	0.00	90.00
5000 - 10000	10.20	10.00	5.00	20.00
10001 - 20000	35.71	20.00	5.00	100.00
More than 20000	29.29	30.00	0.00	70.00

Table 2.1 As a personal estimate about what percentage of your school's online classes utilize zoom?

	Mean	Median	Minimum	Maximum
Entire sample	61.53	67.50	5.00	100.00

Table 2.2 As a personal estimate about what percentage of your school's online classes utilize zoom? Broken out by your college is public or private?

Is your college public or private?	Mean	Median	Minimum	Maximum
Public	59.50	65.00	5.00	100.00
Private	66.00	80.00	5.00	100.00

Table 2.3 As a personal estimate about what percentage of your school's online classes utilize zoom? Broken out by enrollment

Enrollment	Mean	Median	Minimum	Maximum
Less than 4000	48.75	50.00	5.00	100.00
4000 - 9999	54.29	60.00	10.00	95.00
10000 - 20000	78.89	90.00	25.00	100.00
More than 20000	61.13	60.00	5.00	99.00

Table 2.4 As a personal estimate about what percentage of your school's online classes utilize zoom? Broken out by Carnegie Class or type of college

Carnegie Class or type of college	Mean	Median	Minimum	Maximum
Community College	46.00	50.00	5.00	95.00
4-Year College	81.00	80.00	60.00	100.00
MA/PHD Granting	62.50	72.50	5.00	100.00
Research University	66.27	75.00	5.00	99.00

Table 2.5 As a personal estimate about what percentage of your school's online classes utilize zoom? Broken out by tuition, $

Tuition, $	Mean	Median	Minimum	Maximum
Less than 5000	57.78	60.00	5.00	95.00
5000 - 10000	47.14	50.00	15.00	95.00
10001 - 20000	78.63	87.50	5.00	100.00
More than 20000	61.25	77.50	5.00	100.00

Table 3.1 How would you rate the general capabilities of your college faculty in using zoom?

	Overall most are highly proficient	Most are generally proficient	Most are proficient but many need help	Most still need help though some are proficient	Most need a lot of help and only few are proficient
Entire sample	7.89%	42.11%	31.58%	13.16%	5.26%

Table 3.2 How would you rate the general capabilities of your college faculty in using zoom? Broken out by your college is public or private?

Is your college public or private?	Overall most are highly proficient	Most are generally proficient	Most are proficient but many need help	Most still need help though some are proficient	Most need a lot of help and only few are proficient
Public	3.85%	38.46%	34.62%	19.23%	3.85%
Private	16.67%	50.00%	25.00%	0.00%	8.33%

Table 3.3 How would you rate the general capabilities of your college faculty in using zoom? Broken out by enrollment

Enrollment	Overall most are highly proficient	Most are generally proficient	Most are proficient but many need help	Most still need help though some are proficient	Most need a lot of help and only few are proficient
Less than 4000	11.11%	44.44%	33.33%	0.00%	11.11%
4000 - 9999	0.00%	62.50%	12.50%	12.50%	12.50%
10000 - 20000	0.00%	30.00%	50.00%	20.00%	0.00%
More than 20000	18.18%	36.36%	27.27%	18.18%	0.00%

Table 3.4 How would you rate the general capabilities of your college faculty in using zoom? Broken out by Carnegie Class or type of college

Carnegie Class or type of college	Overall most are highly proficient	Most are generally proficient	Most are proficient but many need help	Most still need help though some are proficient	Most need a lot of help and only few are proficient
Community College	0.00%	27.27%	45.45%	18.18%	9.09%
4-Year College	0.00%	50.00%	16.67%	16.67%	16.67%
MA/PHD Granting	25.00%	50.00%	25.00%	0.00%	0.00%
Research University	7.69%	46.15%	30.77%	15.38%	0.00%

Table 3.5 How would you rate the general capabilities of your college faculty in using zoom? Broken out by tuition, $

Tuition, $	Overall most are highly proficient	Most are generally proficient	Most are proficient but many need help	Most still need help though some are proficient	Most need a lot of help and only few are proficient
Less than 5000	10.00%	30.00%	40.00%	10.00%	10.00%
5000 - 10000	0.00%	50.00%	25.00%	12.50%	12.50%
10001 - 20000	9.09%	27.27%	36.36%	27.27%	0.00%
More than 20000	11.11%	66.67%	22.22%	0.00%	0.00%

Table 4.1 How would you rate the general capabilities of your college students in using Zoom?

	Overall most are highly proficient	Most are generally proficient	Most are proficient but many need help	Most still need help though some are proficient	Most need a lot of help and only few are proficient
Entire sample	7.89%	50.00%	26.32%	13.16%	2.63%

Table 4.2 How would you rate the general capabilities of your college students in using Zoom? Broken out by your college is public or private?

Is your college public or private?	Overall most are highly proficient	Most are generally proficient	Most are proficient but many need help	Most still need help though some are proficient	Most need a lot of help and only few are proficient
Public	3.85%	50.00%	26.92%	19.23%	0.00%
Private	16.67%	50.00%	25.00%	0.00%	8.33%

Table 4.3 How would you rate the general capabilities of your college students in using Zoom? Broken out by enrollment

Enrollment	Overall most are highly proficient	Most are generally proficient	Most are proficient but many need help	Most still need help though some are proficient	Most need a lot of help and only few are proficient
Less than 4000	11.11%	55.56%	33.33%	0.00%	0.00%
4000 - 9999	0.00%	62.50%	12.50%	12.50%	12.50%
10000 - 20000	10.00%	50.00%	20.00%	20.00%	0.00%
More than 20000	9.09%	36.36%	36.36%	18.18%	0.00%

Table 4.4 How would you rate the general capabilities of your college students in using Zoom? Broken out by Carnegie Class or type of college

Carnegie Class or type of college	Overall most are highly proficient	Most are generally proficient	Most are proficient but many need help	Most still need help though some are proficient	Most need a lot of help and only few are proficient
Community College	0.00%	36.36%	45.45%	18.18%	0.00%
4-Year College	0.00%	50.00%	16.67%	16.67%	16.67%
MA/PHD Granting	37.50%	50.00%	12.50%	0.00%	0.00%
Research University	0.00%	61.54%	23.08%	15.38%	0.00%

Table 4.5 How would you rate the general capabilities of your college students in using Zoom? Broken out by tuition, $

Tuition, $	Overall most are highly proficient	Most are generally proficient	Most are proficient but many need help	Most still need help though some are proficient	Most need a lot of help and only few are proficient
Less than 5000	0.00%	50.00%	40.00%	10.00%	0.00%
5000 - 10000	12.50%	50.00%	12.50%	12.50%	12.50%
10001 - 20000	9.09%	36.36%	36.36%	18.18%	0.00%
More than 20000	11.11%	66.67%	11.11%	11.11%	0.00%

Table 5 Which of the following features do you find faculty often need help within Zoom? (check all that apply)

Table 5.1.1 Do you find faculty often need any help with setting up passcodes in Zoom?

	No Answer	Yes	No
Entire sample	5.26%	18.42%	76.32%

Table 5.1.2 Do you find faculty often need any help with setting up passcodes in Zoom? Broken out by your college is public or private?

Is your college public or private?	No Answer	Yes	No
Public	0.00%	26.92%	73.08%
Private	16.67%	0.00%	83.33%

Table 5.1.3 Do you find faculty often need any help with setting up passcodes in Zoom? Broken out by enrollment

Enrollment	No Answer	Yes	No
Less than 4000	11.11%	0.00%	88.89%
4000 - 9999	0.00%	12.50%	87.50%
10000 - 20000	0.00%	30.00%	70.00%
More than 20000	9.09%	27.27%	63.64%

Table 5.1.4 Do you find faculty often need any help with setting up passcodes in Zoom? Broken out by Carnegie Class or type of college

Carnegie Class or type of college	No Answer	Yes	No
Community College	0.00%	27.27%	72.73%
4-Year College	0.00%	0.00%	100.00%
MA/PHD Granting	25.00%	0.00%	75.00%
Research University	0.00%	30.77%	69.23%

Table 5.1.5 Do you find faculty often need any help with setting up passcodes in Zoom? Broken out by tuition, $

Tuition, $	No Answer	Yes	No
Less than 5000	0.00%	20.00%	80.00%
5000 - 10000	0.00%	12.50%	87.50%
10001 - 20000	9.09%	27.27%	63.64%
More than 20000	11.11%	11.11%	77.78%

Table 5.2.1 Do you find faculty often need any help with using polls in Zoom?

	No Answer	Yes	No
Entire sample	5.26%	39.47%	55.26%

Table 5.2.2 Do you find faculty often need any help with using polls in Zoom? Broken out by your college is public or private?

Is your college public or private?	No Answer	Yes	No
Public	0.00%	46.15%	53.85%
Private	16.67%	25.00%	58.33%

Table 5.2.3 Do you find faculty often need any help with using polls in Zoom? Broken out by enrollment

Enrollment	No Answer	Yes	No
Less than 4000	11.11%	22.22%	66.67%
4000 - 9999	0.00%	37.50%	62.50%
10000 - 20000	0.00%	40.00%	60.00%
More than 20000	9.09%	54.55%	36.36%

Table 5.2.4 Do you find faculty often need any help with using polls in Zoom? Broken out by Carnegie Class or type of college

Carnegie Class or type of college	No Answer	Yes	No
Community College	0.00%	27.27%	72.73%
4-Year College	0.00%	16.67%	83.33%
MA/PHD Granting	25.00%	50.00%	25.00%
Research University	0.00%	53.85%	46.15%

Table 5.2.5 Do you find faculty often need any help with using polls in Zoom? Broken out by tuition, $

Tuition, $	No Answer	Yes	No
Less than 5000	0.00%	20.00%	80.00%
5000 - 10000	0.00%	62.50%	37.50%
10001 - 20000	9.09%	54.55%	36.36%
More than 20000	11.11%	22.22%	66.67%

Table 5.3.1 Do you find faculty often need any help with recording and transcripts in Zoom?

	No Answer	Yes	No
Entire sample	5.26%	39.47%	55.26%

Table 5.3.2 Do you find faculty often need any help with recording and transcripts in Zoom? Broken out by your college is public or private?

Is your college public or private?	No Answer	Yes	No
Public	0.00%	38.46%	61.54%
Private	16.67%	41.67%	41.67%

Table 5.3.3 Do you find faculty often need any help with recording and transcripts in Zoom? Broken out by enrollment

Enrollment	No Answer	Yes	No
Less than 4000	11.11%	22.22%	66.67%
4000 - 9999	0.00%	37.50%	62.50%
10000 - 20000	0.00%	60.00%	40.00%
More than 20000	9.09%	36.36%	54.55%

Table 5.3.4 Do you find faculty often need any help with recording and transcripts in Zoom? Broken out by Carnegie Class or type of college

Carnegie Class or type of college	No Answer	Yes	No
Community College	0.00%	27.27%	72.73%
4-Year College	0.00%	66.67%	33.33%
MA/PHD Granting	25.00%	37.50%	37.50%
Research University	0.00%	38.46%	61.54%

Table 5.3.5 Do you find faculty often need any help with recording and transcripts in Zoom? Broken out by tuition, $

Tuition, $	No Answer	Yes	No
Less than 5000	0.00%	20.00%	80.00%
5000 - 10000	0.00%	37.50%	62.50%
10001 - 20000	9.09%	63.64%	27.27%
More than 20000	11.11%	33.33%	55.56%

Table 5.4.1 Do you find faculty often need any help with managing student access and admission in Zoom?

	No Answer	Yes	No
Entire sample	5.26%	36.84%	57.89%

Table 5.4.2 Do you find faculty often need any help with managing student access and admission in Zoom? Broken out by your college is public or private?

Is your college public or private?	No Answer	Yes	No
Public	0.00%	42.31%	57.69%
Private	16.67%	25.00%	58.33%

Table 5.4.3 Do you find faculty often need any help with managing student access and admission in Zoom? Broken out by enrollment

Enrollment	No Answer	Yes	No
Less than 4000	11.11%	33.33%	55.56%
4000 - 9999	0.00%	25.00%	75.00%
10000 - 20000	0.00%	40.00%	60.00%
More than 20000	9.09%	45.45%	45.45%

Table 5.4.4 Do you find faculty often need any help with managing student access and admission in Zoom? Broken out by Carnegie Class or type of college

Carnegie Class or type of college	No Answer	Yes	No
Community College	0.00%	63.64%	36.36%
4-Year College	0.00%	33.33%	66.67%
MA/PHD Granting	25.00%	12.50%	62.50%
Research University	0.00%	30.77%	69.23%

Table 5.4.5 Do you find faculty often need any help with managing student access and admission in Zoom? Broken out by tuition, $

Tuition, $	No Answer	Yes	No
Less than 5000	0.00%	50.00%	50.00%
5000 - 10000	0.00%	50.00%	50.00%
10001 - 20000	9.09%	27.27%	63.64%
More than 20000	11.11%	22.22%	66.67%

Table 5.5.1 Do you find faculty often need any help with screen sharing in Zoom?

	No Answer	Yes	No
Entire sample	5.26%	50.00%	44.74%

Table 5.5.2 Do you find faculty often need any help with screen sharing in Zoom? Broken out by your college is public or private?

Is your college public or private?	No Answer	Yes	No
Public	0.00%	57.69%	42.31%
Private	16.67%	33.33%	50.00%

Table 5.5.3 Do you find faculty often need any help with screen sharing in Zoom? Broken out by enrollment

Enrollment	No Answer	Yes	No
Less than 4000	11.11%	44.44%	44.44%
4000 - 9999	0.00%	62.50%	37.50%
10000 - 20000	0.00%	50.00%	50.00%
More than 20000	9.09%	45.45%	45.45%

Table 5.5.4 Do you find faculty often need any help with screen sharing in Zoom? Broken out by Carnegie Class or type of college

Carnegie Class or type of college	No Answer	Yes	No
Community College	0.00%	63.64%	36.36%
4-Year College	0.00%	50.00%	50.00%
MA/PHD Granting	25.00%	25.00%	50.00%
Research University	0.00%	53.85%	46.15%

Table 5.5.5 Do you find faculty often need any help with screen sharing in Zoom? Broken out by tuition, $

Tuition, $	No Answer	Yes	No
Less than 5000	0.00%	50.00%	50.00%
5000 - 10000	0.00%	50.00%	50.00%
10001 - 20000	9.09%	45.45%	45.45%
More than 20000	11.11%	55.56%	33.33%

Table 5.6.1 Do you find faculty often need any help with chat functions in Zoom?

	No Answer	Yes	No
Entire sample	5.26%	18.42%	76.32%

Table 5.6.2 Do you find faculty often need any help with chat functions in Zoom? Broken out by your college is public or private?

Is your college public or private?	No Answer	Yes	No
Public	0.00%	23.08%	76.92%
Private	16.67%	8.33%	75.00%

Table 5.6.3 Do you find faculty often need any help with chat functions in Zoom? Broken out by enrollment

Enrollment	No Answer	Yes	No
Less than 4000	11.11%	0.00%	88.89%
4000 - 9999	0.00%	25.00%	75.00%
10000 - 20000	0.00%	30.00%	70.00%
More than 20000	9.09%	18.18%	72.73%

Table 5.6.4 Do you find faculty often need any help with chat functions in Zoom? Broken out by Carnegie Class or type of college

Carnegie Class or type of college	No Answer	Yes	No
Community College	0.00%	18.18%	81.82%
4-Year College	0.00%	33.33%	66.67%
MA/PHD Granting	25.00%	0.00%	75.00%
Research University	0.00%	23.08%	76.92%

Table 5.6.5 Do you find faculty often need any help with chat functions in Zoom? Broken out by tuition, $

Tuition, $	No Answer	Yes	No
Less than 5000	0.00%	20.00%	80.00%
5000 - 10000	0.00%	12.50%	87.50%
10001 - 20000	9.09%	27.27%	63.64%
More than 20000	11.11%	11.11%	77.78%

Table 5.7.1 Do you find faculty often need any help with whiteboard in Zoom?

	No Answer	Yes	No
Entire sample	5.26%	55.26%	39.47%

Table 5.7.2 Do you find faculty often need any help with whiteboard in Zoom? Broken out by your college is public or private?

Is your college public or private?	No Answer	Yes	No
Public	0.00%	65.38%	34.62%
Private	16.67%	33.33%	50.00%

Table 5.7.3 Do you find faculty often need any help with whiteboard in Zoom? Broken out by enrollment

Enrollment	No Answer	Yes	No
Less than 4000	11.11%	33.33%	55.56%
4000 - 9999	0.00%	50.00%	50.00%
10000 - 20000	0.00%	60.00%	40.00%
More than 20000	9.09%	72.73%	18.18%

Table 5.7.4 Do you find faculty often need any help with whiteboard in Zoom? Broken out by Carnegie Class or type of college

Carnegie Class or type of college	No Answer	Yes	No
Community College	0.00%	54.55%	45.45%
4-Year College	0.00%	50.00%	50.00%
MA/PHD Granting	25.00%	25.00%	50.00%
Research University	0.00%	76.92%	23.08%

Table 5.7.5 Do you find faculty often need any help with whiteboard in Zoom? Broken out by tuition, $

Tuition, $	No Answer	Yes	No
Less than 5000	0.00%	50.00%	50.00%
5000 - 10000	0.00%	62.50%	37.50%
10001 - 20000	9.09%	72.73%	18.18%
More than 20000	11.11%	33.33%	55.56%

Table 5.8.1 Do you find faculty often need any help with calendars in Zoom?

	No Answer	Yes	No
Entire sample	5.26%	10.53%	84.21%

Table 5.8.2 Do you find faculty often need any help with calendars in Zoom? Broken out by your college is public or private?

Is your college public or private?	No Answer	Yes	No
Public	0.00%	11.54%	88.46%
Private	16.67%	8.33%	75.00%

Table 5.8.3 Do you find faculty often need any help with calendars in Zoom? Broken out by enrollment

Enrollment	No Answer	Yes	No
Less than 4000	11.11%	11.11%	77.78%
4000 - 9999	0.00%	0.00%	100.00%
10000 - 20000	0.00%	10.00%	90.00%
More than 20000	9.09%	18.18%	72.73%

Table 5.8.4 Do you find faculty often need any help with calendars in Zoom? Broken out by Carnegie Class or type of college

Carnegie Class or type of college	No Answer	Yes	No
Community College	0.00%	18.18%	81.82%
4-Year College	0.00%	16.67%	83.33%
MA/PHD Granting	25.00%	0.00%	75.00%
Research University	0.00%	7.69%	92.31%

Table 5.8.5 Do you find faculty often need any help with calendars in Zoom? Broken out by tuition, $

Tuition, $	No Answer	Yes	No
Less than 5000	0.00%	20.00%	80.00%
5000 - 10000	0.00%	0.00%	100.00%
10001 - 20000	9.09%	18.18%	72.73%
More than 20000	11.11%	0.00%	88.89%

Table 5.9.1 Do you find faculty often need any help with breakout rooms in Zoom?

	No Answer	Yes	No
Entire sample	5.26%	68.42%	26.32%

Table 5.9.2 Do you find faculty often need any help with breakout rooms in Zoom? Broken out by your college is public or private?

Is your college public or private?	No Answer	Yes	No
Public	0.00%	73.08%	26.92%
Private	16.67%	58.33%	25.00%

Table 5.9.3 Do you find faculty often need any help with breakout rooms in Zoom? Broken out by enrollment

Enrollment	No Answer	Yes	No
Less than 4000	11.11%	55.56%	33.33%
4000 - 9999	0.00%	75.00%	25.00%
10000 - 20000	0.00%	80.00%	20.00%
More than 20000	9.09%	63.64%	27.27%

Table 5.9.4 Do you find faculty often need any help with breakout rooms in Zoom? Broken out by Carnegie Class or type of college

Carnegie Class or type of college	No Answer	Yes	No
Community College	0.00%	81.82%	18.18%
4-Year College	0.00%	50.00%	50.00%
MA/PHD Granting	25.00%	75.00%	0.00%
Research University	0.00%	61.54%	38.46%

Table 5.9.5 Do you find faculty often need any help with breakout rooms in Zoom? Broken out by tuition, $

Tuition, $	No Answer	Yes	No
Less than 5000	0.00%	80.00%	20.00%
5000 - 10000	0.00%	75.00%	25.00%
10001 - 20000	9.09%	63.64%	27.27%
More than 20000	11.11%	55.56%	33.33%

Table 5.10.1 Do you find faculty often need any help with security features in Zoom?

	No Answer	Yes	No
Entire sample	5.26%	44.74%	50.00%

Table 5.10.2 Do you find faculty often need any help with security features in Zoom? Broken out by your college is public or private?

Is your college public or private?	No Answer	Yes	No
Public	0.00%	46.15%	53.85%
Private	16.67%	41.67%	41.67%

Table 5.10.3 Do you find faculty often need any help with security features in Zoom? Broken out by enrollment

Enrollment	No Answer	Yes	No
Less than 4000	11.11%	22.22%	66.67%
4000 - 9999	0.00%	50.00%	50.00%
10000 - 20000	0.00%	70.00%	30.00%
More than 20000	9.09%	36.36%	54.55%

Table 5.10.4 Do you find faculty often need any help with security features in Zoom? Broken out by Carnegie Class or type of college

Carnegie Class or type of college	No Answer	Yes	No
Community College	0.00%	45.45%	54.55%
4-Year College	0.00%	16.67%	83.33%
MA/PHD Granting	25.00%	50.00%	25.00%
Research University	0.00%	53.85%	46.15%

Table 5.10.5 Do you find faculty often need any help with security features in Zoom? Broken out by tuition, $

Tuition, $	No Answer	Yes	No
Less than 5000	0.00%	50.00%	50.00%
5000 - 10000	0.00%	37.50%	62.50%
10001 - 20000	9.09%	45.45%	45.45%
More than 20000	11.11%	44.44%	44.44%

Other:

1. I have not used many of the features
2. accessibility
3. Accessibility for persons with disabilities
4. Classroom management: monitoring the text chat while teaching. I have long-advocated that every instructor needs a technologically astute Graduate student.

Table 6.1 Has your school developed videos or tutorials to help in learning Zoom?

	No Answer	Yes	No	No but in the process of making them
Entire sample	13.16%	42.11%	42.11%	2.63%

Table 6.2 Has your school developed videos or tutorials to help in learning Zoom? Broken out by your college is public or private?

Is your college public or private?	No Answer	Yes	No	No but in the process of making them
Public	15.38%	50.00%	30.77%	3.85%
Private	8.33%	25.00%	66.67%	0.00%

Table 6.3 Has your school developed videos or tutorials to help in learning Zoom? Broken out by enrollment

Enrollment	No Answer	Yes	No	No but in the process of making them
Less than 4000	11.11%	33.33%	55.56%	0.00%
4000 - 9999	12.50%	62.50%	25.00%	0.00%
10000 - 20000	30.00%	30.00%	40.00%	0.00%
More than 20000	0.00%	45.45%	45.45%	9.09%

Table 6.4 Has your school developed videos or tutorials to help in learning Zoom? Broken out by Carnegie Class or type of college

Carnegie Class or type of college	No Answer	Yes	No	No but in the process of making them
Community College	9.09%	63.64%	27.27%	0.00%
4-Year College	33.33%	33.33%	33.33%	0.00%
MA/PHD Granting	0.00%	25.00%	75.00%	0.00%
Research University	15.38%	38.46%	38.46%	7.69%

Table 6.5 Has your school developed videos or tutorials to help in learning Zoom? Broken out by tuition, $

Tuition, $	No Answer	Yes	No	No but in the process of making them
Less than 5000	0.00%	60.00%	40.00%	0.00%
5000 - 10000	12.50%	50.00%	37.50%	0.00%
10001 - 20000	27.27%	18.18%	45.45%	9.09%
More than 20000	11.11%	44.44%	44.44%	0.00%

Please write for your peers a short thirty second address on exactly what to avoid in Zoom and exactly what do to use it to its greatest potential. Broken out by Carnegie Class or type of college

Community College

1) Avoid using any new feature without testing it first. Only use features that suit the purpose of instructional support. If you use it in real-time make sure learners are actively engaged.
2) When an instructor visits the Zoom room with students for a "tour" of tutoring options, and that screen is shared via the LMS, be aware that sound does not hop from interface to interface. While students can see the room, they can't hear the tutor's voice over the LMS interface.
3) When using Zoom, try to practice in a non-classroom setting before bringing it to students. Also, make sure you can use the mute functions and that students have some basic zoom netiquette.
4) What to do: forget about the tool at first. Think of the tasks you need to do or that you would be doing in your class (group discussions and projects), formative assessment and comprehension checks, shared document production on the white board... then devise a way to use the Zoom tools available to achieve those same goals.
5) Don't be intimidated about using Zoom. Mistakes will be made, but just keep going. It's called learning.
6) Doing PowerPoints in zoom is challenging when you have dual screens - getting it to slideshow view background noise when people don't mute is irritating Video/webcam challenges are difficult to troubleshoot Greatest potential: easy to use; relatively stable performance-wise
7) Don't just lecture at them and don't try to replicate your campus classroom. Use breakout rooms, polls, and chats paired with pre-class assignments to up the engagement.

4-Year College

1) Don't go public, put passcode on (else get zoom bombed), use screen sharing
2) Avoid overuse of Zoom in class. Two-three hour sessions are too long for many students and it interferes with their learning.

MA/PHD Granting

1) Planning, faculty need to plan and test what they plan on doing in advance of class time if it is a new process for them.
2) Two keys to success in Zoom calls are using the waiting room function and sending reminders 10 minutes before the meeting starts.
3) Do not try to recreate the classroom lecture. Use it to answer questions, clarify confusion and generally check on people.
4) We use the schools Single Sign-on system to avoid zoom bombing. Train student employees prior to start of classes.
5) Zoom is not an enterprise application just yet but has been adopted by many faculty in Spring 2020. The breakout feature of zoom meetings allows faculty the flexibility of group assignments without technology challenges. It is important to lock meetings and enable the waiting room to prevent unwanted participants in the meeting session.
6) I would simply direct them to the Zoom online documentation at https://zoom.us/docs/doc/Comprehensive%20Guide%20to%20Educating%20Through%20Zoom.pdf.

Research University

1) Don't: leave everyone to unmute/mute themselves. Practice ahead of time! Do: provide the one-tap phone number for your local area (i.e. help people avoid long distance fees by mistake)
2) use it to teach, advise, department meetings
3) Ensure that you pay attention to the security settings
4) Learn functions ahead of time, before they are need. Make sure security features are enabled (passwords, etc.). Join sessions early just in case there are connection issues. For some applications, a webinar will be better than a meeting.
5) Avoid just lecturing but take advantage of interactive features such as breakout rooms and chat.
6) Think about if you need to deliver information (-> provide a video or text for students to work on their own time) or stimulate exchanges (use Zoom and interactive teaching strategies such as clicker questions, break out rooms, etc.). Keep it short!
7) Use the security features. Do not publicly publish the Zoom link.
8) Almost all of the downsides are overblown and something of a moral panic. The ease with which Zoom meetings can be incorporated into the normal LMS workflow - in a whole slew of ways - and it's fundamental dependability and flexibility make it far superior than the other options we have (BBCollaborate, Teams, etc.)

9) Use a waiting room to intentionally admit participants, bring your chat window back after screen share, and teach people to use mute.
10) what not to do...

What has been your most effective tutorial or other tool to bring instructors and students up to speed in Zoom? Broken out by Carnegie Class or type of college

Community College

1) Zoom YouTube channel videos on using breakout rooms.
2) ??
3) It is most effective to let faculty "play" in the program, talk through what it is they'd like to do with students, and then work through solutions. As far as students, they should know the basic equipment they need, preferable a computer with a camera and reliable internet.
4) I have been doing faculty development on online teaching practice, but I use Zoom as I would hope the instructors would use it in their classes, securely and using the tools to have them feedback, collaborate, and share. I have had several instructors say, "hey, I'm going to use that in my class." So, generally, model best practices and creativity in Zoom use myself and have the instructors experience it as students.
5) There are many on California ConferZoom Web site. I just recently sent a video on Booking an Appointment with me.
6) We have made our own videos for faculty and students showing how to set up meetings and access meetings - short videos with key features are the best
7) an open discussion among faculty on their tips and experiences

4-Year College

1) Our 'Manage Participants' tutorial
2) short quick videos
3) Breakout rooms

MA/PHD Granting

1) Zoom Help pages/videos
2) The Zoom website tutorials and job aids.
3) Demonstrating it while holding meetings.
4) Zoom videos and screen shots
5) We have not begun training faculty on Zoom. We plan to create video tutorials along with work aids. These two tutorial methods meet the need of

the majority of the faculty. One on one training sessions will be available for faculty as well.

6) (see above)

Research University

1) practice
2) Youtube and Zoom web site
3) https://support.zoom.us/hc/en-us/articles/206618765-Zoom-Video-Tutorials
4) We have a brief how-to tutorial on how to use Zoom, coupled with Zoom-provided material. We also have a more advanced tutorial on how to set up Zoom to let students record presentations, which has been successful.
5) Zoom resources
6) Documented "use cases" presenting approaches for exercise sessions, meeting with TAs etc. of other instructors at the institution
7) Live training demonstration with written tutorials as reference
8) The beauty is that it barely needs any.
9) my office has not been doing the majority of zoom support. The ITS office has been doing it so I am not sure what they would say.
10) firm

Which programs or applications do you think best complement Zoom for college classes? Broken out by Carnegie Class or type of college Broken out by Carnegie Class or type of college

Community College

1) Because you can record sessions, it works well with most other programs and applications.
2) GoBoard?
3) Screen sharing, hand raising function, break out rooms.
4) Synchronous: Kahoot for engagement...the game goes on just like in class.
5) Canvas
6) Blackboard works fine; we have little integration between zoom and other applications
7) not sure
8) Canvas

4-Year College

1) LMS - we use Canvas. There are a few other integrations, like with Teams or Slack that make our IT Services staff a bit more adaptive.
2) We use Canvas, Microsoft Teams too

MA/PHD Granting

1) n/a
2) The whiteboard feature.
3) PowerPoint or other presentation software that you can share.
4) Canvas, Micro-soft office suite
5) Cisco WebEx Meetings. We currently have this application adopted enterprise wide. If Microsoft Teams increases the number of participants that can be seen in the meeting, it will be a competitive product.
6) I am a long-time user of Blackboard Learn and Blackboard Collaborate. Zoom is fine for web conferencing, but it does not have the tools necessary for a virtual classroom.

Research University

1) breakout rooms google docs
2) group activities, lecture classes
3) LMS
4) For the most part, Zoom works well by itself, though it is only part of the tool set. For example, our LMS is used to coordinate activities, some of which

involve Zoom. For certain courses, we also recommend a tablet so instructors can annotate the whiteboard in Zoom.

5) Turning Point for better polling; the LMS
6) Clickers (i.e. SpeakUp, TurningTechnologies, etc.) LMS (i.e. Moodle) to provide students with structured information (dates, links, etc.)
7) The learning management system
8) I found I can integrate it with Canvas almost seamlessly.
9) slack, google docs

Once the pandemic concludes and either through a vaccine, treatment, superb contact tracing or herd immunity, college life returns to something approaching normal, how do you think colleges will utilize Zoom? Broken out by Carnegie Class or type of college

Community College

1) I hope they will continue using it as a synchronous tool.
2) We will likely still host online tutoring sessions and study groups for those who want them.
3) I think colleges can use Zoom in hybrid courses (partially on-line and partially face to face). Also, it can be used in fully on-line sections and for students to collaborate remotely.
4) Could be an option for students away on long-term absences Where I live, many of us are expats. It would be interesting to see if Zoom would be allowed for faculty or students who are unable to be in-country temporarily during the semester.
5) I think it will fall back a bit, but still will be a useful tool. Screen share is a great feature to help students with their issues.
6) Continue using for meetings and coursework
7) much less but for limited use for office hours, sick students, more meetings, and back up plans
8) I believe many faculty will continue to use Zoom in both fully online courses and as an enhancement to face to face courses such as for office hours.
9) no

4-Year College

1) Large groups. It's become apparent that having 300 faculty and staff in a room is far less effective than having them all watch it in their offices and making it available on demand for those that miss the meeting.
2) Same as it has now, more meetings over zoom than in person.
3) Colleges should continue to use Zoom in the classroom for online synchronous meetings

MA/PHD Granting

1) More synchronous zoom courses.
2) As a tool for connecting with students for office hours and other functions.
3) I hope they use them for meetings, they seem to go faster when on Zoom than in face to face.
4) I think there should be a part of Zoom in each class for 21st century learning.

5) Post pandemic, Zoom will be used to allow athletes to view class while traveling for their sport, colleagues to conduct meetings with other colleagues that may be located on the other side of campus, and faculty will use the tool for office hours and advising with students. These are just a few ways I see this tool continuing to be used.
6) I think it
7) hybrid flexible models

Research University

1) maybe for some classes
2) advising, office hours, hybrid instruction
3) My fear is that they go back to teaching as usual. Lectures
4) I think a number of face-to-face meetings will be replaced by Zoom, just because it saves time and physical spaces.
5) I think it will be used for some flipped teaching, as well as office hours and tutorials.
6) I am interpreting this question to mean "use Zoom for teaching." No, not much. I think that colleges will use more asynchronous ways of delivering information online, but that synchronous interactions will come back on campus.
7) Maybe for remote, instead of in-person, office hours; interviews with people not based in the same local area;
8) I suspect we will use it for a lot of meetings and maybe some advising and grading and group work. Especially the latter. It's good for professional students, in particular, to get used to working on teams but it's a real burden for some to be able to be in the same place/time as others so their comfort and skill gained here with Zoom will go far.
9) Online meetings for staff will largely continue to be high, and Zoom will likely continue as an option with in-person classes.
10) I think utilization will decrease quite a bit. Most faculty want to go back to teaching the way they did before all this. There will likely be a few who see applications for it going forward but not many.
11) more

Table 7.1 Over the past two months how serious have the security problems been with your college's use of Zoom?

	No Answer	Extremely concerning	Very concerning	Somewhat concerning	Not so concerning	Not at all concerning
Entire sample	18.42%	10.53%	2.63%	15.79%	36.84%	15.79%

Table 7.2 Over the past two months how serious have the security problems been with your college's use of Zoom? Broken out by your college is public or private?

Is your college public or private?	No Answer	Extremely concerning	Very concerning	Somewhat concerning	Not so concerning	Not at all concerning
Public	19.23%	7.69%	3.85%	15.38%	34.62%	19.23%
Private	16.67%	16.67%	0.00%	16.67%	41.67%	8.33%

Table 7.3 Over the past two months how serious have the security problems been with your college's use of Zoom? Broken out by enrollment

Enrollment	No Answer	Extremely concerning	Very concerning	Somewhat concerning	Not so concerning	Not at all concerning
Less than 4000	11.11%	11.11%	0.00%	11.11%	55.56%	11.11%
4000 - 9999	12.50%	25.00%	0.00%	12.50%	50.00%	0.00%
10000 - 20000	30.00%	10.00%	10.00%	10.00%	30.00%	10.00%
More than 20000	18.18%	0.00%	0.00%	27.27%	18.18%	36.36%

Table 7.4 Over the past two months how serious have the security problems been with your college's use of Zoom? Broken out by Carnegie Class or type of college

Carnegie Class or type of college	No Answer	Extremely concerning	Very concerning	Somewhat concerning	Not so concerning	Not at all concerning
Community College	9.09%	18.18%	0.00%	9.09%	45.45%	18.18%
4-Year College	33.33%	16.67%	0.00%	16.67%	33.33%	0.00%
MA/PHD Granting	12.50%	12.50%	0.00%	12.50%	50.00%	12.50%
Research University	23.08%	0.00%	7.69%	23.08%	23.08%	23.08%

Table 7.5 Over the past two months how serious have the security problems been with your college's use of Zoom? Broken out by tuition, $

Tuition, $	No Answer	Extremely concerning	Very concerning	Somewhat concerning	Not so concerning	Not at all concerning
Less than 5000	0.00%	20.00%	0.00%	10.00%	30.00%	40.00%
5000 - 10000	12.50%	12.50%	0.00%	12.50%	62.50%	0.00%
10001 - 20000	45.45%	0.00%	9.09%	36.36%	9.09%	0.00%
More than 20000	11.11%	11.11%	0.00%	0.00%	55.56%	22.22%

What advice do you give faculty and students about protecting Zoom sessions from intrusion or hacking? Broken out by Carnegie Class or type of college

Community College

1) I have not had anyone looking to disrupt my community college classes. However, I decided to use the waiting room feature.
2) We use a form to collect student data about their visit before they come to the Zoom room. That way, we can verify their enrollment while they are in the Zoom waiting room if there is any question about their identity.
3) Carefully set up zoom links and make sure attendees are manually admitted into the meeting.
4) Limit sharing and invitations to Zoom meetings. Remove anyone who does not show their video or have a registered name on the screen. Allow screensharing only when required.
5) Let's help each other. I think the "Waiting" really helps with this issue.
6) password protect, no public link posting, require registration, keep option to remove and report participants to Zoom
7) Be sure to be familiar with all the security features of Zoom and activate them as appropriate. If an intruder enters your Zoom room, be sure to know how to remove them. Consider appointing a cohost to assist.
8) use the security controls with zoom (e.g., waiting room, force muting, limiting the ability to screenshare)

4-Year College

1) Use the security tools and training that we've given you. 9/10 times the complaints come from someone that did zero planning and publicly shared a link with no password through a listserv.
2) Password protect, waiting room
3) Enable the waiting room and set a pass code to enter meetings.

MA/PHD Granting

1) Don't publish links publicly
2) Use the waiting room feature and let the system generate new identification numbers for each meeting instead using the same one.
3) Put the link to the meeting behind a password protected wall, or on a course website. Also, use the waiting rooms and don't share the screen without knowing who you are sharing it with.
4) Go through single sign on
5) We have informed them to lock the meeting room, enable the waiting room and set passwords for meetings.

6) Use a password and a waiting room, lock the meeting when you no longer expect attendees

Research University

1) use a password and waiting room, look over the participants list
2) none, has not been a problem
3) Follow the security settings
4) Zoom has always had security features, it's just that many opted not to use them, so we had to remind users about these features. We also suggested that for some cases, webinars are better than meetings, because they allow you more audience control.
5) Use it through the LMS.
6) Whenever possible (& certainly for large classes) have a TA or responsible student designated as cohost. This person can be attentive to disruptions/issues and use Zoom function to deal with it. This leaves the main teacher free to concentrate on the interactions with students
7) Use a password, waiting room, authentication. Do not post the link on social media platforms.
8) Don't be a bozo. Use a password. Don't post invites on Facebook, etc. It's really not a big deal. Lots of the issues that the so-called experts were warning about are things that had already been resolved or things that Zoom had in common with other platforms. The amount of productivity lost because of MSTeams impervious UI and klunky experience makes Zoom problems look inconsequential in the extreme. Convenience and ease of use matter.
9) Use waiting room for intentionally admitting participants
10) use passwords for sessions mostly and don't make session links public. The idea of having them use waiting rooms and then allowing people in is beyond most of them to understand why or how to do that.

What are some of the very basic things that faculty and students tend not to know about Zoom? Broken out by Carnegie Class or type of college

Community College

1) It is easy to use across devices. You do not need to have your camera on.
2) Breakout rooms - that we can throw people in one to work together without one of the staff having to be in there with them.
3) Zoom netiquette and mute function.
4) We use Zoom through our LMS. Most faculty don't realize they have a more extensive dashboard through zoom.us
5) Virtual Backgrounds. Live captioning needs real improvement. Not as good as Google Meetings. Maybe License that from Google?
6) How to share screen; mute people and manage participants
7) that you can record, that people often forget about mute options, that you can show video, you can share screens and presentations or files and do student presentations, you can record smaller concept focused sections by starting and stopping the recording throughout a long session.
8) How to integrate it with Outlook. How to configure the security features.
9) to have the mute on when listening in a group setting

4-Year College

1) -Basic accounts are time limited but pro accounts are not. -The limit for the number of people on a call - it's much higher for both Basic and Pro than most would think.
2) The Zoom mobile app - they forget that it is an excellent tool to connect to a meeting if there are technical issues with the main computer.

MA/PHD Granting

1) Sharing screens and especially video
2) How to share their screens or whiteboard.
3) Break out rooms. I wish they could be set up ahead of time with participants, but alas they cannot.
4) How to use whiteboard and polling
5) Since we have not adopted it enterprise wide, I do not have much information to provide. Overall, the majority of faculty work with the tool without help from IT.

Research University

1) you can raise your hand. Host can mute/unmute people.
2) whiteboard, polling
3) The security settings
4) It's a synchronous tool and requires a stable internet connection. Just because Netflix works great at your house, Zoom may not because it cannot cache like Netflix does in the background to account for network disruptions.
5) How to mitigate bandwidth issues, how to use hand raising feature.
6) To activate file share in chat -> useful for students to work on exercises To get students to change their name in the participant list to the name of the group they want to join -> then "dispatch" TA can send them there
7) Features like whiteboard, breakout rooms, and polls.
8) The same things they don't know about most of the tools they use. And most of these are in the category of things they should not need to know.
9) how to turn video on/off, screen sharing, granting permissions for others to become host, proper use of chat feature (who you are chatting with), how to raise your hand to speak is a big one - even I can't always find it!

Made in the USA
Middletown, DE
03 November 2026

36494702R00031